MathStart
UNDERSTANDING HALVES

Give Me Half !

by Stuart J. Murphy · **illustrated by G. Brian Karas**

HarperCollinsPublishers

LEVEL
2

To Randy and Kristin—
who have each done their share
of saying "Give Me Half"
—S.J.M.

To Elizabeth and Andrew
—G.B.K.

The illustrations in this book were done with acrylics, gouache, and pencils on Strathmore.

For more information about the MathStart series, please write to
HarperCollins Children's Books, 10 East 53rd Street, New York, NY 10022.

Bugs incorporated in the MathStart series design were painted by Jon Buller.

HarperCollins®, ▰®, and MathStart™ are trademarks of HarperCollins Publishers Inc.

Give Me Half!
Text copyright © 1996 by Stuart J. Murphy
Illustrations copyright © 1996 by G. Brian Karas.
Printed in the U.S.A. All rights reserved.

Library of Congress Cataloging-in-Publication Data
Murphy, Stuart J.
   Give me half! / by Stuart J. Murphy ; illustrated by G. Brian Karas.
      p.   cm. (MathStart. Level 2)
   Summary: Introduces the concept of halves using a simple rhyming story about a brother
and sister who do not want to share their food.
   ISBN 0-06-025873-X. — ISBN 0-06-025874-8 (lib. bdg.)
   ISBN 0-06-446701-5 (pbk.)
   1. Fractions—Juvenile literature.   2. Division—Juvenile literature.   [1. Fractions.]
I. Karas, G. Brian, ill.   II. Title.   III. Series.
QA117.M87   1996   95-19617
513.2'6—dc20   CIP
                                                                                       AC

Typography by Elynn Cohen
5   6   7   8   9   10
❖

Give Me Half !

I have one whole pizza . . . and it's all for me!

I'm going to get some pizza—just you wait and see.

I know you want some pizza, Sis.
You only get one slice.

You'd better give me more than that.
Why can't you be nice?

7

You have to share the pizza. It must be split in two.
The pieces should be cut the same for each of you.

and make

1 half pizza and 1 half pizza make 1 whole pizza.

$\frac{1}{2}$ and $\frac{1}{2}$ is 1

$\frac{1}{2} + \frac{1}{2} = 1$

What is that behind her back?
The last can of juice, I think.

If he takes some juice from me,
I won't have much to drink.

I gave you half of mine, so you must share yours, too.

I'll give you just a sip, but not until I'm through.

Split the juice in half. Again, you have to share.
And when you pour it out,
make sure that you pour fair.

and make

1 half can of juice and 1 half can of juice
make 1 whole can of juice.

$\frac{1}{2}$ and $\frac{1}{2}$ is 1

$\frac{1}{2} + \frac{1}{2} = 1$

I know she has some cupcakes.
I saw her with a pack.

I'm going to hide my cupcakes
and save them for a snack.

Hey, what's that on your chair?
You'd better give me some!

I'm going to eat them both myself
and leave you just a crumb.

Your cupcakes must be shared—and do I need repeat?
You both get half the pack.
And don't you two dare cheat!

and make

1 cupcake and 1 cupcake make a pack of 2 cupcakes.

1 and 1 is 2

1 + 1 = 2

1 is ½ of 2

I have a stack of cookies—
and you get just one bite.

You'd better give me half that stack,
or else I'll start a fight!

Hey . . . WAIT!

Too late!

There are cookies on the floor.
There's pizza everywhere.

There's juice spilled on the table
and sticky stuff under my chair.

We're going to be in trouble.
We've made a great big mess!
But I have made no more than you,
and you have made no less.

We'd better each clean half.
There's so much work to do.
We'll be done in half the time . . .

if Buddy helps us too.

FOR ADULTS AND KIDS

I f you would like to have fun with the math concepts presented in *Give Me Half!*, here are a few suggestions:

- Read the story with the child and describe what is going on in each picture.

- Ask questions throughout the story, such as "What happens when you share one whole pizza with another person?" and "How much pizza will you get?" and "How should the pizza be cut if it is to be shared fairly?"

- Encourage the child to tell the story using the math vocabulary: "Half," "Whole," "Share," etc. Introduce the word "divide" by saying that each item is "divided equally."

- Gather pieces of paper in a variety of sizes and shapes and work together to find different ways to fold the pieces in half.

- Together, draw some pizzas, cups of juice, cupcakes, and cookies, or draw your family meal or an imaginary picnic. Then cut or make lines on the drawings to show how they should be divided when sharing with another person.

- Look at things in the real world—whole items such as pies or tables or rooms, liquids such as milk or water, and groups of things such as grapes or candy or pillows—and ask how they can be divided in half. Draw pictures of these halves.

(Continued on next page)

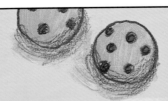

Following are some activities that will help you extend the concepts presented in *Give Me Half!* into a child's everyday life.

Cooking: Identify halves in a recipe: one half of one cup or of one stick of butter, halves of fruits and vegetables, etc. Slice food items of varying shapes into halves and then share them.

Nature: Collect lots of different leaves and use a marker to show how they divide in half.

Games: Find things around the house that demonstrate halves—two small shoes that together are about the same length as one long shoe, shirts and pants that can be folded in half, etc.

The following stories include some of the same concepts that are presented in *Give Me Half!*:

- THE DOORBELL RANG by Pat Hutchins

- GATOR PIE by Louise Mathews

- THE HALF-BIRTHDAY PARTY by Charlotte Pomerantz